Slugs

by Anthony D. Fredericks
photographs by Gerry Ellis

Lerner Publications Company • Minneapolis, Minnesota

To Lori Leckrone, whose passion for slugs was the inspiration for this book
—ADF

Thanks to our series consultant, Sharyn Fenwick, elementary science/math specialist. Mrs. Fenwick was the winner of the National Science Teachers Association 1991 Distinguished Teaching Award. She also was the recipient of the Presidential Award for Excellence in Math and Science Teaching, representing the state of Minnesota at the elementary level in 1992.

Additional photographs are reproduced with the permission of the following sources: Tom Stack and Associates: © Dave B. Fleetham, p. 8; © James P. Rowan, pp. 9, 42; Visuals Unlimited: (© R. Calentine) p. 16, (© Frank T. Awbrey) p. 24; © Scott Camazine/The National Audubon Society Collection/Photo Researchers, Inc., pp. 28, 39; © Yuko Sato, p. 32 (inset); © Frank Balthis, p. 33; Animals Animals/ © Bromhall, N., p. 35; © Peter Ward/Bruce Coleman, Inc., p. 36; Illustration by John Erste, p. 19.

Early Bird Nature Books were conceptualized
by Ruth Berman and designed by Steve Foley.
Series editor is Joelle Riley.

Lerner Publications Company
A Division of Lerner Publishing Group
241 First Avenue North
Minneapolis, MN 55401 U.S.A.

Website address: www.lernerbooks.com

Library of Congress Cataloging-in-Publication Data

Fredericks, Anthony D.
    Slugs / by Anthony D. Fredericks ; photographs by Gerry Ellis.
      p. cm — (Early bird nature books)
    Includes index.
    Summary: Describes the physical characteristics,
  habitat, and behavior of slugs, slimy creatures
  that spend their lives crawling on their stomachs.
    ISBN 0–8225–3041–4 (alk. paper)
    1. Slugs (Mollusks)—Juvenile literature. [1. Slugs
  (Mollusks)] I. Ellis, Gerry, ill. II. Title. III. Series.
  QL430.4.F84 2000
  594'.38–dc21   JJ 594.38   98–49295

Manufactured in the United States of America
1 2 3 4 5 6 – JR – 05 04 03 02 01 00

Contents

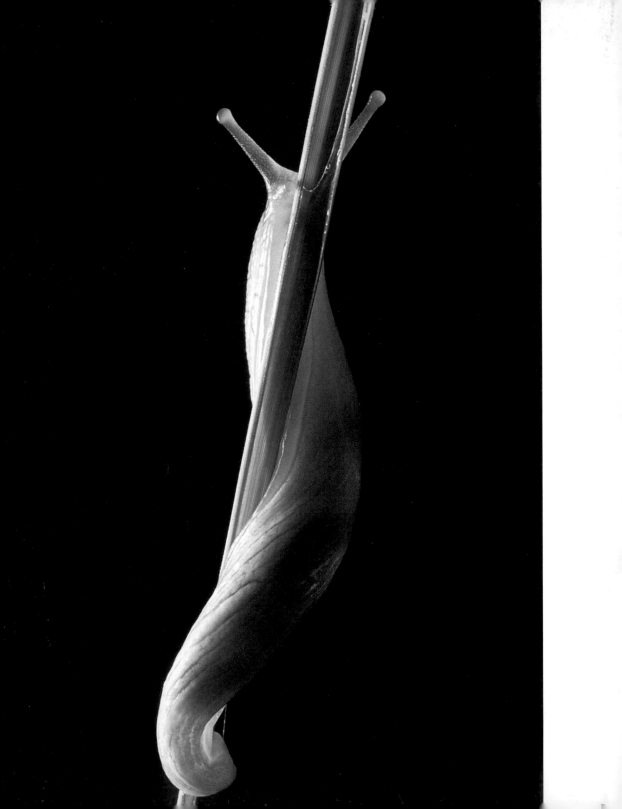

Be a Word Detective

Can you find these words as you read about the slug's life? Be a detective and try to figure out what they mean. You can turn to the glossary on page 46 for help.

cannibals   mollusk   predators
gastropods   mucus   radula
herbivorous   omnivorous   tentacles
mantle

Chapter 1

There are thousands of different kinds of slugs. How many feet does a slug have?

A Snail without a Shell

 Imagine an animal that has only one foot. Imagine that the animal's foot is really its

belly. Imagine that the animal moves along the ground on its belly. Then, imagine that animal covered with icky, gooey slime. The animal you have imagined is real. It is a slug.

These slugs are called banana slugs. Banana slugs can be white, black, yellow, tan, or other colors.

Octopuses are mollusks.

Slugs belong to a large group of animals called mollusks (MOLL-uhsks). Snails, clams, squids, and octopuses are mollusks too. Mollusks are animals with a soft body. They have no skeleton in their body.

Some mollusks are called gastropods (GAS-truh-pods). Gastropods are animals who crawl on their belly. Slugs and snails are gastropods.

The word gastropod *means "bellyfoot."*

Many gastropods have shells. A snail has a shell that covers its body. A slug doesn't have a shell on the outside. But some species, or kinds, of slugs have a tiny shell hidden inside their body.

Snails are much like slugs. But snails have a hard shell covering part of their body.

Slugs breathe through a hole in their skin. The hole is on the right side of a slug's body.

There is a patch of thick skin on a slug's back. This patch of skin is called the mantle. The main part of a slug's body is called the foot.

Slugs' eyes are at the end of their tentacles. Each eye is round, like a ball (inset).

Slugs have two pairs of tentacles (TEN-tuh-kuhls) on their head. Tentacles are thin, flexible stalks. The top pair of tentacles is longer than the bottom pair. The top pair of tentacles have eyes at their tips. But a slug cannot see the way you can. It can only tell light from dark. Slugs use their bottom pair of tentacles to help them find food. This is like when you use your nose to tell that there are fresh-baked cookies nearby. A slug also uses its bottom pair of tentacles to feel objects.

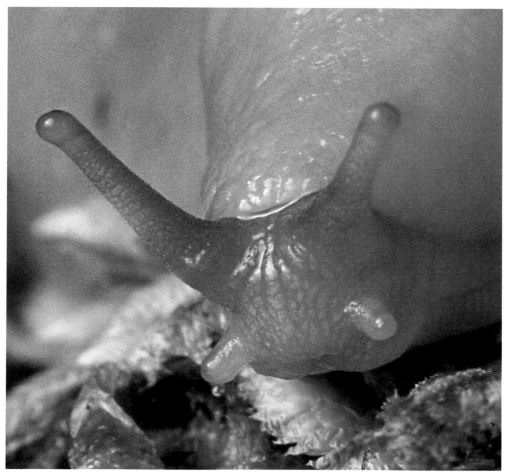

A slug uses its bottom pair of tentacles for tasting.

The largest slugs are sea slugs. Sea slugs are usually brightly colored.

There are two main groups of slugs. One group is the sea slugs. The other group is the land slugs.

Sea slugs are found in every ocean. Some sea slugs crawl on the ocean bottom. Some swim through the water.

Land slugs live on the land. Some species of land slugs live in cold parts of the world. Other land slugs live in hot, wet areas. And a few species live in dry places. Almost anywhere you go in the world, you'll find land slugs.

Many kinds of slugs live in damp places like this rain forest.

Chapter 2

Slugs are always covered in slime. What is this slime called?

Slime Patrol

Have you ever seen a slug crawl along the ground? If you have, you've probably seen the trail it left behind. The trail is made of gooey slime called mucus (MYOO-kuhs). Mucus is a

thick liquid. It is made by the slug's body.
A slug's body is covered with mucus.

Most slugs make two different kinds of
mucus. One kind of mucus is sticky. The other
kind is slippery.

*One kind of mucus is sticky. Slugs use sticky mucus
to climb trees.*

A slug's foot is wide. It has many muscles.

Mucus helps a slug glide over rough
surfaces and sharp objects. The mucus is like
a thick coat. It protects the slug from hurting
its foot.

A slug moves by making ripples along its foot. These ripples move from the back to the front of the slug's foot. They look like waves in the ocean.

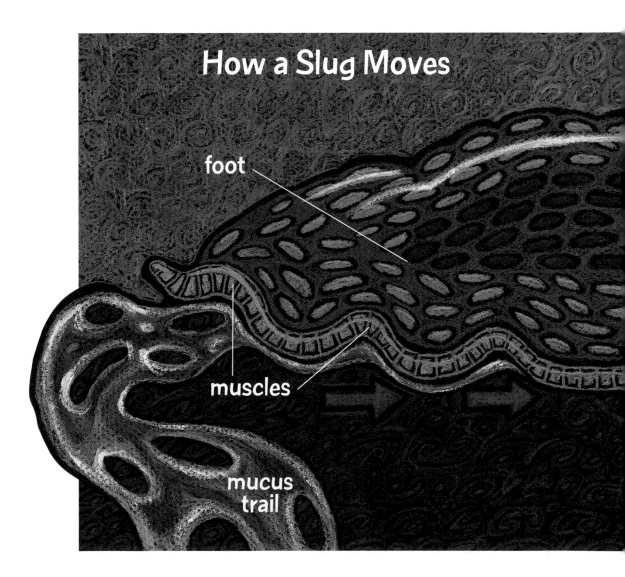

How a Slug Moves

foot

muscles

mucus trail

Slugs spend much of their time in damp places such as under these leaves.

All animals need water in their body to stay healthy. We lose water from our body when we sweat. If we lose too much water, we become sick. Slugs also must keep water from leaving their body. So slugs spend most of their time in damp places.

If slugs go out in the sun, they will dry up. Slugs hide from the sun in moist soil. They also hide under plants or under rocks.

At night the sun is gone. The air is cool and damp. Then slugs can leave their hiding places.

Most slugs live both above the ground and underground. But some kinds of slugs spend their whole life under the ground.

At night, most slugs go out into the open air. When they go out, there is no moist soil to keep them wet. But the mucus covering a slug's body keeps water inside.

Nighttime is when many slugs go out into the open air.

Banana slugs live in places that get lots of rain.

Sometimes, the air becomes too dry or too hot. Then a slug digs deep into moist ground. It covers its body with extra mucus. Then it curls up and waits for rain. The slug's body can soak up rainwater from the ground.

In very cold weather, some slugs sleep under the ground. They sleep until the weather is warmer.

Mucus also helps protect slugs from cold weather. If the weather is too cold, slugs dig into the soil. They cover themselves with a thick coat of mucus. They wait until the weather is warmer. Slugs may stay buried for several months. When it is warm enough, the slugs wriggle up out of the ground. Then they begin looking for something to eat.

The scientific name of the great gray garden slug is Limax maximus.

*This is a great gray garden slug. It is using a string
of mucus to get down from a leaf.*

Some species of slugs use mucus to get
down from trees. These slugs can climb trees
30 feet or higher. This is taller than most
houses. When the slugs want to come down,

26

The slug is on another leaf. It can crawl again.

they make a thick string of mucus. They use the mucus like a rope to come down from the tree.

A slug's mucus may seem icky. But it is an important part of how a slug lives.

Chapter 3

Some slugs eat animals. But other slugs eat only plants. What are slugs who eat only plants called?

Munch, Munch, Munch

Some species of slugs are herbivorous (her-BIH-ver-uhs). Herbivorous slugs eat only plants. They eat mostly leaves, stems, and roots. These plant parts contain lots of water. Other slugs are omnivorous (ahm-NIH-ver-uhs). Omnivorous slugs eat both plants and animals.

European black slugs eat plants from people's gardens.

Garbage makes a good meal for these slugs.

Omnivorous slugs eat almost anything.
They eat garbage, dead animals, manure,

and rotting plants. Some omnivorous slugs are predators (PREH-duh-turz). Predators are animals who kill and eat other animals. Predatory slugs eat animals such as earthworms, centipedes, and snails.

Some slugs eat snails.

A slug's radula has rows of tiny teeth. Inset: This is a close-up of a radula.

Slugs eat with their radula (RAD-juh-luh). The radula is like a tongue. It has many tiny teeth on its surface. Some species of slugs have more than 20,000 teeth on their radula. Other slugs have fewer teeth. When a slug wants to

A slug spends most of its time resting, eating, or looking for food.

eat something, it pushes its radula out of its mouth. The slug rubs its radula back and forth on the food. The radula's teeth scrape off small pieces of food. Then the slug sucks the pieces of food into its mouth.

Chapter 4

Most kinds of slugs lay 3 to 50 eggs at a time. Where do slugs lay their eggs?

Slug Babies

Slugs lay eggs when the weather is warm. Many kinds of slugs lay their eggs in July or August. Slugs lay their eggs in damp, shady places. The eggs are usually stuck together in groups. After a slug lays its eggs, it goes away.

Slug eggs take several weeks or months to hatch. Baby slugs may hatch in the fall. Or they may not hatch until the next spring. The baby slugs look just like adult slugs. But they are much smaller than adults. Baby slugs are usually less than a quarter of an inch long. This is about as long as the nail of your pinky finger.

This slug is hatching from its shell.

Slugs usually lay their eggs beneath a piece of wood or in a hole in the ground. These are newly hatched slugs.

Life is hard for baby slugs. They have to take care of themselves. Their parents are not around to help them. Baby slugs must find their own food. They must protect themselves from enemies. It takes about six months for baby slugs to grow up.

This slug is small. The largest slugs grow to be over 10 inches long.

Chapter 5

Some animals don't like mucus, so they do not eat slugs. But mice eat slugs. What other animal enemies do slugs have?

Dangers

Slugs have many animal enemies. Frogs, mice, badgers, and raccoons are some of the animals who eat slugs. Baby slugs have even more enemies. They are also eaten by centipedes and beetles.

Slugs even need to stay away from other slugs. Some species of slugs are cannibals (KA-nuh-bulz). Cannibals are animals who eat

their own kind. A cannibal slug eats other
slugs. It does this if it cannot find other food.
First, the cannibal slug eats the mucus from
another slug. Then it chews on the skin.
Finally, it eats the slug's insides.

Great gray garden slugs eat other slugs.

Slugs eat fruits that are close to the ground. Tomatoes and strawberries grow close to the ground.

Some slugs have human enemies too. Many people dislike herbivorous slugs. That's because some of these slugs eat the plants in people's gardens. These slugs often eat tomato plants. And they eat other fruit plants that are near the ground. A favorite food of these slugs is seedlings. Seedlings are baby plants.

This seedling will grow into a bean plant. Sometimes slugs eat bean seedlings.

Slugs live one to six years.

Most slugs are not pests. They help the earth. Slug droppings help plants grow in the soil. Some slugs eat dead plants and animals.

This helps clean up the earth. Many people think slugs are harmful. But slugs are important in nature.

Slugs help to keep the earth healthy.

On Sharing a Book

As you know, adults greatly influence a child's attitude toward reading. When a child sees you read, or when you share a book with a child, you're sending a message that reading is important. Show the child that reading a book together is important to you. Find a comfortable, quiet place. Turn off the television and limit other distractions, such as telephone calls.

Be prepared to start slowly. Take turns reading parts of this book. Stop and talk about what you're reading. Talk about the photographs. You may find that much of the shared time is spent discussing just a few pages. This discussion time is valuable for both of you, so don't move through the book too quickly. If the child begins to lose interest, stop reading. Continue sharing the book at another time. When you do pick up the book again, be sure to revisit the parts you have already read. Most importantly, enjoy the book!

Be a Vocabulary Detective

You will find a word list on page 5. Words selected for this list are important to the understanding of the topic of this book. Encourage the child to be a word detective and search for the words as you read the book together. Talk about what the words mean and how they are used in the sentence. Do any of these words have more than one meaning? You will find these words defined in a glossary on page 46.

What about Questions?

Use questions to make sure the child understands the information in this book. Here are some suggestions:

What did this paragraph tell us? What does this picture show? What do you think we'll learn about next? Can slugs live in your backyard? Why/Why not? Why do slugs make mucus? Why do slugs hide in damp places? How does a slug move? How does a slug eat? What things are dangerous to slugs? For what do slugs use their tentacles? What is your favorite part of this book? Why?

If the child has questions, don't hesitate to respond with questions of your own such as: What do *you* think? Why? What is it that you don't know? If the child can't remember certain facts, turn to the index.

Introducing the Index

The index is an important learning tool. It helps readers get information quickly without searching throughout the whole book. Turn to the index on page 47. Choose an entry, such as *eating,* and ask the child to use the index to find out what slugs eat. Repeat this exercise with as many entries as you like. Ask the child to point out the differences between an index and a glossary. (The index helps readers find information quickly, while the glossary tells readers what words mean.)

All the World in Metric

Although our monetary system is in metric units (based on multiples of 10), the United States is one of the few countries in the world that does not use the metric system of measurement. Here are some conversion activities you and the child can do using a calculator:

WHEN YOU KNOW:	MULTIPLY BY:	TO FIND:
miles	1.609	kilometers
feet	0.3048	meters
inches	2.54	centimeters
pounds	0.454	kilograms

Activities

Make up a story about a slug. Be sure to include information from this book. Draw or paint pictures to illustrate your story.

Find a slug in your yard or in a park. What color is it? Is it spotted? Does it have stripes? Find the slug's mantle. Find the slug's breathing hole. Touch the slug's mucus. How would you describe the mucus? Draw a picture of the slug, then take the picture to the library and try to find out what kind of slug it is.

The fastest slugs can travel .025 mile per hour. If you were to travel at that speed, you'd move about 26 feet in one minute. Lie flat on your belly and see how far you can crawl in one minute. Make sure your stomach is always touching the ground!

Glossary

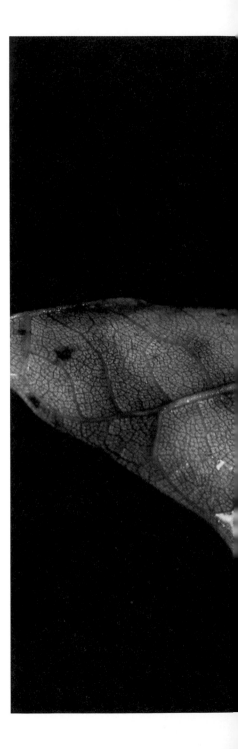

cannibals (KA-nuh-bulz)—animals who eat their own kind

gastropods (GAS-truh-pods)—animals who crawl on their belly

herbivorous (her-BIH-ver-uhs)—eating only plants

mantle—the patch of thick skin on a slug's back

mollusks (MOLL-uhsks)—animals who have soft bodies. Octopuses, clams, and slugs are all mollusks.

mucus (MYOO-kuhs)—the thick slime that a slug makes

omnivorous (ahm-NIH-ver-uhs)—eating both plants and animals

predators (PREH-duh-turz)—animals who kill and eat other animals

radula (RAD-juh-luh)—a slug's raspy tongue

tentacles (TEN-tuh-kuhls)—thin, flexible stalks

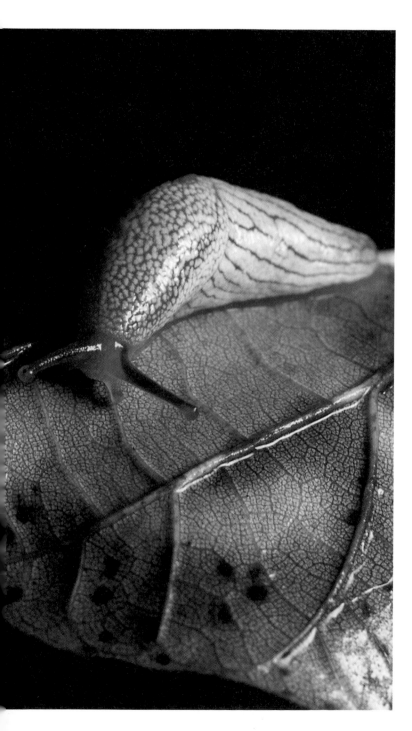

Index

Pages listed in **bold** type refer to photographs.

About the Author

Anthony D. Fredericks is a former elementary classroom teacher and reading specialist. In his position as a teacher educator at York College in York, Pennsylvania, he teaches courses in science and language arts education. He has written nearly 300 articles and more than four dozen books, including college textbooks, teacher resource books, and children's books. Several of his books have won special awards and citations. He maintains a website for elementary teachers (www.afredericks.com) that includes up-to-the-minute ideas in science education. In his free time he enjoys traveling, snorkeling, hiking, and exploring.

About the Photographer

Gerry Ellis has explored the world as a professional photographer and naturalist for nearly two decades. His images of wildlife and natural landscapes have won him many awards, including several honors in the BBC Wildlife Photographer of the Year competition. Among his many publications are the Lerner Publishing Group titles *Hippos, Cheetahs, Rhinos,* and *African Elephants.* Mr. Ellis lives in Portland, Oregon.

The Early Bird Nature Books Series

African Elephants	Horses	Sandhill Cranes
Alligators	Jellyfish	Scorpions
Ants	Manatees	Sea Lions
Apple Trees	Moose	Sea Turtles
Bobcats	Mountain Goats	Slugs
Brown Bears	Mountain Gorillas	Swans
Cats	Peacocks	Tarantulas
Cockroaches	Penguins	Tigers
Cougars	Polar Bears	Venus Flytraps
Crayfish	Popcorn Plants	Vultures
Dandelions	Prairie Dogs	Walruses
Dolphins	Rats	Whales
Giant Sequoia Trees	Red-Eyed Tree Frogs	White-Tailed Deer
Herons	Saguaro Cactus	Wild Turkeys